THE POETRY OF IRON

The Poetry of Iron

Walter the Educator™

SKB

Silent King Books a WhichHead Imprint

Copyright © 2023 by Walter the Educator™

All rights reserved. No part of this book may be reproduced in any manner whatsoever without written permission except in the case of brief quotations embodied in critical articles and reviews.

First Printing, 2023

Disclaimer
This book is a literary work; poems are not about specific persons, locations, situations, and/or circumstances unless mentioned in a historical context. This book is for entertainment and informational purposes only. The author and publisher offer this information without warranties expressed or implied. No matter the grounds, neither the author nor the publisher will be accountable for any losses, injuries, or other damages caused by the reader's use of this book. The use of this book acknowledges an understanding and acceptance of this disclaimer.

"Earning a degree in chemistry changed my life!"
- Walter the Educator

dedicated to all the chemistry lovers, like myself, across the world

CONTENTS

Dedication v

Why I Created This Book? 1

One - Earth's Design 2

Two - Symbol Of Might 4

Three - Testament To Endurance 6

Four - Magnetic Allure 8

Five - Bold And Bright 10

Six - Modern Machines 12

Seven - Unwavering And Strong 14

Eight - Symbol Of Power 16

Nine - Desires Conspire 18

Ten - Never Tire 20

Eleven - Dances And Sings 22

Twelve - Honor Iron 24

Thirteen - Strength Unmatched 26

Fourteen - Rust And Steel 28

Fifteen - Countless Memoirs 30

Sixteen - Iron's Might 32

Seventeen - Ancient Hands 34

Eighteen - Weaves Through History 36

Nineteen - Age To Age 38

Twenty - Shaped And Tamed 40

Twenty-One - Proud And Tall 42

Twenty-Two - Blood Of Wars 44

Twenty-Three - Towering Structures 45

Twenty-Four - Legacy Divine 47

Twenty-Five - Year After Year 49

Twenty-Six - Bends Or Breaks 51

Twenty-Seven - Grit And Grace 53

Twenty-Eight - Spirit Of The Land 55

Twenty-Nine - Craftsmen And Kings 57

Thirty - Human Spirit And Might 59

Thirty-One - Shapes And Molds 61

Thirty-Two - Span The Divide 63

Thirty-Three - Face Of Time	. . . .	65
Thirty-Four - Centuries Untold	. . . .	67
Thirty-Five - Tempestuous Gales	. . . .	69
Thirty-Six - Dawn To The Dusk	. . . .	71
About The Author	. . . .	73

WHY I CREATED THIS BOOK?

Creating a poetry book about the chemical element of Iron was a fascinating endeavor because it allows for the exploration of the various facets of iron in a creative and artistic way. Iron, with its rich history, symbolism, and significance in science and industry, provides a wealth of material for poetic exploration. I can delve into the physical properties of iron, its role in human history and culture, and its metaphorical implications to create a diverse and thought-provoking collection of poems. This unique theme can offer a fresh perspective and engage readers in a captivating journey through science and art.

ONE

EARTH'S DESIGN

In the heart of the earth, deep and wrought,
Lies a metal strong, with battles fought.
Iron, the stalwart, steadfast and true,
Forged in flames, its strength imbued.
    From the core, it rises high and wide,
A guardian, a protector, by our side.
In the hands of craftsmen, it takes form,
Shaping empires, weathering every storm.
    Its presence felt in the blood we share,
Binding life with an unyielding flair.
Magnetized whispers in the northern dance,
Iron, the essence of our earthly romance.
    Rust may claim its surface, a silent plea,
Yet its spirit endures, unyielding and free.

An anchor, a sword, a bridge to the past,
Iron, the legacy that forever will last.
    So let us honor this element so grand,
A symbol of strength, in every land.
From the forge of creation to the end of time,
Iron, the soul of the earth's design.

TWO

SYMBOL OF MIGHT

In the heart of the earth, where flames once danced,
Lies a metal strong, with an ancient romance.
Forged in the crucible of fire and heat,
Iron emerges, steadfast and complete.
    Guardian of empires, protector in wars,
Iron stands tall, amidst earthly scars.
Magnetic whispers in its core,
Drawing us near, forevermore.
    In the blood that flows, a bond so deep,
Iron sustains life, a promise to keep.
Rust may claim its earthly form,
But its spirit endures, through calm and storm.
    A legacy of strength, through ages untold,
Iron's story is written in empires of old.

Honor its presence, a symbol of might,
A legacy that will endure, day and night.

THREE

TESTAMENT TO ENDURANCE

In the heart of the earth, where flames once danced,
Lies a metal of ancient romance.
Forged in the crucible of fire and stone,
Iron, the stalwart guardian, stands alone.
    Beneath the soil, in darkness deep,
Lies a legacy that mortals keep.
A symbol of strength, enduring and true,
Iron, the protector, watches over you.
    From the swords of warriors to the wheels that turn,
Iron's presence is felt, its legacy will never burn.
In the forge of time, it stood the test,
A testament to endurance, it is the best.
    Majestic in its simplicity, steadfast in its might,
Iron, the silent sentinel, shines in the light.

A bond unbroken, through ages untold,
Iron, the steadfast soul, never grows old.
    So, let us honor this metal so grand,
For it is the foundation on which civilizations stand.
In its quiet resilience, we find our own,
Iron, the enduring spirit, forever known.

FOUR

MAGNETIC ALLURE

In the heart of the earth, a mighty force,
Lies a metal of strength, unwavering and coarse.
Iron, the guardian of civilization's birth,
Forged in fire, embodying resilience and worth.
    Silent sentinel, standing tall and true,
Enduring the ages, a timeless virtue.
From towering skyscrapers to swords of old,
Iron shapes empires, steadfast and bold.
    In the furnace of creation, it takes its form,
A symbol of endurance, in calm and storm.
Binding nations, with its unyielding might,
Iron stands as a testament to human insight.
    In the hands of craftsmen, it becomes art,
A foundation of progress, a beating heart.

Magnetic allure, drawing souls near,
Iron, the essence of strength, resolute and clear.
   Through the annals of time, its legacy prevails,
A steadfast companion, as history unveils.
So let us honor this metal, noble and grand,
For in iron's embrace, civilizations stand.

FIVE

BOLD AND BRIGHT

In the heart of earth's deep embrace,
Lies a metal, steadfast and strong,
Forged in fire, with unyielding grace,
Iron, ancient, where it belongs.
    From the dawn of time, it stood the test,
A guardian of civilizations past,
Shaping empires at its behest,
Majestic, enduring, built to last.
    In the hands of warriors, it gleamed,
A sword, a shield, a symbol of might,
On battlefields where destinies teemed,
Iron bore witness to honor and fight.
    In towering structures, it found its place,
Binding nations with bridges of might,

A testament to human resilience and grace,
Iron, the foundation of progress and light.
    So let us honor this metal so true,
As we forge ahead, our dreams in sight,
For in Iron's strength, we find the clue,
To build a future, bold and bright.

SIX

MODERN MACHINES

In the heart of the earth, Iron lies,
Forged in fire, under ancient skies.
A metal of might, steadfast and true,
Enduring the ages, a warrior through.
    In the furnace of time, Iron was born,
Unyielding and strong, from dusk till morn.
From swords to plows, it shaped our fate,
A silent guardian, steadfast and great.
    In armor and chains, it stood the test,
Defender of realms, in battle's crest.
In towering structures, it rose on high,
A symbol of power, reaching the sky.
    From the anvils of old to modern machines,
Iron's legacy runs through history's scenes.
In the hands of craftsmen, it bends and molds,
A testament to might, as history unfolds.

So, sing of Iron, noble and bold,
A pillar of strength, a story untold.
In the annals of time, its tale shall endure,
A metal of legends, steadfast and pure.

SEVEN

UNWAVERING AND STRONG

In the heart of the earth, a silent sentinel lies,
Forged in the fiery depths where chaos dances and dies.
Iron, ancient and enduring, whispers tales untold,
Of empires risen and fallen, of mysteries yet to unfold.
From the weapons of war to the plow in the field,
Iron shapes the destiny of humanity, its power revealed.
It binds the very fabric of progress and human insight,
A steadfast companion in the day and in the night.
Through the ages it has stood, unwavering and strong,
Witness to triumph and tragedy, to right and to wrong.
In its unyielding embrace, civilizations have grown,

And in its steadfast resolve, seeds of greatness are sown.
Oh, Iron, noble and true, your presence shapes our fate,
Your enduring spirit, a symbol of power and might so great.
As we look to the future, your legacy will endure,
A testament to resilience, steadfast and pure.

EIGHT

SYMBOL OF POWER

In the heart of the earth, Iron sleeps,
Forged in fire, where darkness weeps.
A metal mighty, steadfast and true,
Binding civilizations, old and new.
    From swords of old to towering spires,
Iron breathes life into human desires.
In battles fierce, it stood the test,
A symbol of power, unmatched, and blessed.
    Through time and tide, it does endure,
A foundation strong, steadfast, and pure.
In the hands of craftsmen, it takes new form,
Shaping the world, through calm and storm.
    Oh, Iron, your legacy profound,
In every corner of the world, you're found.

A silent witness to history's sway,
You stand unwavering, come what may.
    So here's to Iron, unyielding and bold,
A story untold, yet always retold.
In the dance of progress, you play your part,
Forever engraved in the human heart.

NINE

DESIRES CONSPIRE

In the heart of Earth's embrace, Iron lies,
Forged in fires of ancient skies.
A metal mighty, strong and true,
Its tale unfolds in shades of blue.
    From the blood that runs so deep,
To the swords that silence sleep,
Iron weaves through history's thread,
A symbol of both life and dread.
    In battles fierce, it takes its stand,
Shaping kingdoms, shaping land.
The anvils ring, the hammers beat,
As Iron shapes both triumph and defeat.
    In towering structures it finds its place,
Binding cities with its embrace.

Bridging chasms, reaching high,
Iron touches the endless sky.
    Resilient and unyielding, it stands the test,
A testament to time, it's truly blessed.
From ancient times to modern days,
Iron's strength forever stays.
    So let us cherish this metal bold,
A story of endurance, still untold.
For in its veins, our desires conspire,
Iron, the element that fuels our fire.

TEN

NEVER TIRE

In the heart of the earth, where fire meets stone,
Lies a metal of strength, with secrets unknown.
Iron, the guardian of ancient lore,
Shaper of empires, from distant days of yore.
    Beneath the forge of heaven's eternal flame,
It yields to the will of those who seek its acclaim.
From sword to plow, it bends to human demand,
Crafting the tools that have shaped the land.
    In battles fierce, it sang its deadly song,
Yet in peace, it built the world, sturdy and strong.
From towering cities to bridges that span,
Iron stands steadfast, the mark of mortal man.
    A witness to time, it rusts and yet endures,
A silent sentinel, steadfast and sure.

From the Iron Age to the modern age's call,
It whispers of history, standing proud and tall.
    So let us honor this metal so true,
For in its story, our own is woven too.
From the depths of the earth to the heights we aspire,
Iron, the eternal, shall never tire.

ELEVEN

DANCES AND SINGS

In the heart of the earth, where fire meets stone,
Lies a metal so mighty, a king on his throne.
Iron, they call it, with strength to behold,
Forged in the furnace, in darkness and cold.

From ancient swords to towering spires,
Iron shapes civilizations, ignites their fires.
It bends to our will, a servant so true,
In war and in peace, it carries us through.

In the hands of the blacksmith, it dances and sings,
Crafting the tools that the laborer brings.
It whispers of conquest, of triumph and might,
Yet weaves through our history in shadows of night.

It's the anchor, the chain, the bridge and the wheel,
A silent companion, steadfast as steel.

In the clash of the battle, it roars with the fight,
Yet in moments of peace, it reflects the sunlight.
　So here's to the iron, so sturdy and bold,
A story of ages, in legends untold.
It's the backbone of progress, the heart of the flame,
Iron, forever, shall etch out its name.

TWELVE

HONOR IRON

In the heart of the earth, where fire and stone entwine,
Lies a metal of strength, a symbol of mankind's design.
Iron, forged in the furnace of time's relentless gaze,
Shaped our destiny in countless, unyielding ways.

From ancient swords to towering spires that scrape the sky,
Iron has been our companion, never asking why.
In battles fierce, it lent its might to warriors bold,
And in peaceful fields, it built the bridges that unite and hold.

Through ages past, it stood unwavering, unbroken and true,
A testament to resilience, in all that humans pursue.

In the hands of artisans, it took on forms diverse,
Melding with other elements, a dance of endless verse.
   From the plow that tills the soil to the engines that roar,
Iron powers our progress, opening every door.
In its silent strength, we find our own reflected grace,
A reminder of the human spirit, in every time and place.
   So, let us honor Iron, in all its steadfast might,
For it's the backbone of our story, in the day and in the night.
A companion through the ages, a witness to our toil,
Iron, the unyielding metal, beneath the earth's rich soil.

THIRTEEN

STRENGTH UNMATCHED

In the heart of ancient fires, Iron was born,
Forged in earth's fiery embrace, a metal to adorn.
From the blades of warriors to the wheels that turn,
Iron's strength and might, forever will it yearn.
    In the blacksmith's hands, it takes on new form,
Crafted into tools, a testament to its norm.
Binding nations with railways, a network so vast,
Iron's legacy endures, from present to the past.
    In towering structures, it stands tall and proud,
A symbol of progress, amidst the bustling crowd.
Bridging the chasms with steadfast support,
Iron weaves tales of triumph, in every fort.
    In the heat of battle, it shields and it arms,
A protector of kingdoms, with unwavering charms.

Mighty and unyielding, in the forge of war,
Iron shapes destinies, like never before.
    From the anvils of time, its story unfolds,
A testament to resilience, as history beholds.
In the fabric of civilizations, its mark deeply etched,
Iron, the timeless ally, with strength unmatched.

FOURTEEN

RUST AND STEEL

In the heart of the earth, where fires glow,
Lies a metal that kingdoms sought to know.
Iron, the forger of empires grand,
Shaping swords and plows with its steady hand.
    Bridging vast gaps with its steadfast might,
Binding nations in its unyielding sight.
From the depths of mines to the sky's embrace,
Iron endures, leaving a timeless trace.
    In fortress walls and towering spires,
It stands resilient against time's dire.
A steadfast ally in war and in peace,
Iron's strength and grace will never cease.
    Yet in its veins, a flexibility lies,
Bending to the will of human cries.

Molded into tools, machines, and art,
Iron serves mankind in every part.
    From the clang of hammers to the roar of trains,
Iron fuels progress through its enduring reigns.
A legacy written in rust and steel,
Iron's story is timeless, vibrant, and real.

FIFTEEN

COUNTLESS MEMOIRS

Iron, the silent sentinel of time,
Forged in the heart of earth's fiery prime.
A metal of strength, unyielding and bold,
Shaping the world as history unfolds.
    In the heat of the forge, it bends to our will,
Crafting the tools that our dreams fulfill.
From the plow to the sword, it serves without fail,
A loyal companion, in every tale.
    In battles it stood, unwavering and true,
A shield, a weapon, for the chosen few.
Through ages of conquest, it bore the scars,
A witness to triumphs, and countless memoirs.
    From towering skyscrapers to bridges that span,
It binds and connects, as only it can.

A foundation of progress, steadfast and sure,
Enduring the ages, timeless and pure.
    Iron, oh iron, in all forms and hues,
A symbol of might, in the hands we use.
So here's to the metal that never grows old,
Iron, the legend, forever untold.

SIXTEEN

IRON'S MIGHT

In the heart of mountains, Iron lies,
Forged in earth, where fire flies,
Ancient companion of humankind,
In battles fought, and peace enshrined.
    From plows that till the soil so deep,
To swords that clash, and cities keep,
Iron, steadfast, in its many forms,
Shapes the land, and weathers storms.
    A bridge across the river wide,
Binding nations, side by side,
Iron spans the chasm's breadth,
Uniting hearts, conquering death.
    In industry, it fuels the fire,
A catalyst of progress, never to tire,

Machinery hums, with Iron's might,
Turning darkness into radiant light.
    In the crucible of war and peace,
Iron endures, it will not cease,
A testament to human will,
Forged in fire, unyielding still.
    So here's to Iron, steadfast and true,
A mirror of the human spirit too,
In strength and endurance, it stands,
A cornerstone of civilizations' hands.

SEVENTEEN

ANCIENT HANDS

In the heart of the earth, where fire meets stone,
Lies a metal so mighty, a king on his throne.
Iron, the forger of empires untold,
In the furnace of time, its story unfolds.
 From the first sword drawn in ancient hands,
To the rails that unite far distant lands,
Iron, unyielding, in battles it sings,
Forging the future on its tempered wings.
 Bridging the chasms, spanning the seas,
Binding the world with its silent decrees.
In towers and spires, it reaches the skies,
A testament to the strength that never dies.
 Flexing and bending to the will of mankind,
In plows and in hammers, its purpose defined.
Loyal companion in peace and in strife,
Shaping our history with unyielding life.

So here's to the iron, steadfast and true,
In the blood of the earth, in the work that we do.
A symbol of progress, a force to admire,
Iron, the element that will never tire.

EIGHTEEN

WEAVES THROUGH HISTORY

In the heart of the earth, where fire and stone entwine,
Lies a metal of strength, a force divine.
Iron, forged in the furnace of time,
A symbol of progress, an ancient paradigm.
From the first sparks of industry to the modern age,
Iron has fueled our evolution, page by page.
It binds nations, shapes civilizations tall,
A bridge of strength, standing through every fall.
In the forge's glow, it takes on many forms,
A sword, a plow, a ship that braves the storms.
It bends and twists, yet never breaks,
A testament to resilience, whatever it takes.
In battles fought and victories won,
Iron has seen the rise and setting of the sun.

It weaves through history, an unyielding thread,
A silent witness to the blood that's shed.
    From the railways that span continents wide,
To the skyscrapers that pierce the sky with pride,
Iron stands as a testament to human might,
A silent guardian, through day and night.
    So here's to the iron, steadfast and true,
A symbol of endurance, in all that we do.
In its unyielding strength, we find our own,
A legacy of iron, through ages, it's shown.

NINETEEN

AGE TO AGE

In the heart of the earth, Iron lies,
Forged in fire, under ancient skies.
A bridge between past and present days,
In its silent strength, history stays.
    From swords of old to towering spires,
It fuels progress, fuels our desires.
In battles fierce, it's a steadfast shield,
In peace and war, its power revealed.
    Iron, the cornerstone of our might,
Binding nations, forging what's right.
An unyielding force, through time it endures,
In its veins, the world's story ensures.
    So let it be known, from age to age,
Iron's legacy on history's page.

A testament to its unyielding will,
Shaping the future, standing still.

TWENTY

SHAPED AND TAMED

In the heart of the earth, Iron sleeps,
Forged in fire, where silence weeps.
A steadfast soul, unyielding might,
In war and peace, a guiding light.
    From ancient swords to modern steeds,
Iron endures and intercedes.
It spans the ages, a bridge of time,
Connecting nations, a bond divine.
    In the furnace of progress, it takes its form,
A symbol of strength, through calm and storm.
Witness to history, it stands tall and true,
Shaping the world, in ways old and new.
    In battles fierce, it never bends,
A loyal companion, on which life depends.

It whispers tales of rise and fall,
A witness to humanity's call.
   In the hands of men, it's been shaped and tamed,
A legacy etched, forever acclaimed.
From the earth it rises, and to it returns,
Iron, the link, where history yearns.

TWENTY-ONE

PROUD AND TALL

Behold the mighty iron, steadfast and true,
Forged in the heart of ancient stars, a bond so strong and new.
Its atoms weave a tale of ages past,
Where empires rose and fell, their echoes destined to last.

From swords of old to towering skyscrapers high,
Iron's resolve unyielding, it touches the earth and sky.
In battles fierce, it stood the test of time,
A silent witness to humanity's climb.

In rail and road, it binds the lands as one,
Connecting distant shores beneath the blazing sun.
Through iron's veins, the pulse of progress beats,
A testament to human spirit that never retreats.

So let us honor iron, in all its might and grace,
A symbol of endurance, a silent guardian of our race.

In every beam, in every nail, it tells a story untold,
Of civilizations born and legends yet to unfold.
    As time marches on, and ages come and go,
Iron remains unyielding, in its silent, stoic glow.
A tribute to the past, a bridge to the future's call,
Iron, the eternal witness, standing proud and tall.

TWENTY-TWO

BLOOD OF WARS

In the heart of the earth, where flames once danced,
Lies a metal mighty, in strength entranced.
Iron, the silent witness of time's grand parade,
In battles and progress, its mark never to fade.
From ancient swords to towering skyscrapers tall,
Iron's story unfolds, standing proud and enthralled.
It binds nations together, spanning land and sea,
Forging paths of progress, shaping history.
In the fiery forge, where hammers strike and ring,
Iron takes its form, a symbol of enduring.
It bears the weight of empires, the blood of wars,
Yet remains unyielding, steadfast evermore.
Through the ages it endures, unwavering and true,
A testament to resilience, in all that it went through.
Oh, iron, steadfast guardian of humanity's tale,
Your strength and endurance shall forever prevail.

TWENTY-THREE

TOWERING STRUCTURES

In the heart of Earth, forged in fire's embrace,
Lies a metal of strength, with an enduring grace.
Iron, the silent witness to humanity's story,
From ancient empires to modern glory.

It binds nations in bridges, tall and grand,
Connecting distant shores, uniting the land.
A symbol of resilience, steadfast and true,
Forging a path for the old and the new.

In battles it stood, a shield and a sword,
Unyielding and fierce, an ancient lord.
It shaped the world with its unwavering hand,
Crafting history, where civilizations stand.

From towering structures to wheels that spin,
Iron's mark is etched deep within.

A testament to time, it does not fade,
A legacy of strength, in every shade.
    So here's to iron, steadfast and bold,
A tale of endurance, from days of old.
In the veins of the Earth, and the spirit of men,
Iron endures, now and forever again.

TWENTY-FOUR

LEGACY DIVINE

In the heart of the earth, where fire and metal entwine,
Lies a steadfast spirit, unyielding and divine.
Iron, the silent witness to wars and peace,
Forged in ancient fires, its strength will never cease.
From the swords of warriors to the wheels of progress,
Iron has shaped our world, a symbol of duress.
Bridging nations and cultures with its unyielding grace,
It stands tall and unbroken, in every time and place.
In the clang of hammers and the roar of machines,
Iron sings a song of resilience, woven in our genes.
Through the ravages of time, it stands unwavering and strong,
A testament to human spirit, enduring all along.

From the anvils of history to the towers of tomorrow,
Iron binds our story, in joy and also sorrow.
So let us honor this metal, this beacon of might,
For it connects our past and future, in the eternal fight.

In the veins of the earth, where fire and metal entwine,
Lies a spirit unyielding, a legacy divine.
Iron, the guardian of progress and might,
Shall endure through the ages, in the morning's golden light.

TWENTY-FIVE

YEAR AFTER YEAR

In the forge of time, iron stands,
A steadfast symbol, shaped by hands.
From molten core to towering spire,
It fuels progress, ignites the fire.
    Mighty swords and chains it yields,
Binding nations, on battlefields.
Yet in its strength, a silent grace,
It builds the future, marks our place.
    In the heart of stars, its birth begun,
A cosmic dance, its story spun.
Through ages past, its presence known,
An ancient ally, unyielding and honed.
    From steam and steel to bridges tall,
In every triumph, it stands through all.

A testament to human will,
An iron spirit, unbroken still.
    So let us honor, with hearts sincere,
This metal marvel, year after year.
For in its atoms, we find our own,
Resilient, enduring, like iron shown.

TWENTY-SIX

BENDS OR BREAKS

In the heart of the earth, forged in fire,
Lies a metal that all can admire.
Iron, steadfast and true, unyielding might,
Shaping history with its enduring light.
    From ancient swords to towering towers,
Iron's strength has withstood time's powers.
Through battles fought and empires risen,
It echoes the resilience of the human vision.
    In the hands of artisans, it transforms,
Crafting tools and machines in all forms.
From locomotives to the structures we raise,
Iron whispers the tale of our endless praise.
    It binds the past to the future's call,
A bridge of progress, standing tall.

In its atoms, a story of human endeavor,
A testament to our spirit that will last forever.
    So let us honor this element so grand,
For it holds the dreams of our collective hand.
Iron, the metal that never bends or breaks,
Symbol of progress, for all our sakes.

TWENTY-SEVEN

GRIT AND GRACE

In realms of fire and earth, a steadfast force,
Iron, the binding link in nature's course.
Forged in stars, from cosmic dust it came,
To shape the world and bear the weight of fame.
    From ancient swords to modern towers tall,
Iron weaves a story, standing strong through all.
Bridging distant lands with rails of steel,
It whispers tales of progress, of dreams that are real.
    In the heart of the forge, it finds its birth,
A symbol of strength, resilience and worth.
Rust may gnaw and time may wear,
Yet iron endures, beyond compare.
    In the hands of artisans, it takes new form,
Crafting tools and art, weathering any storm.

A silent sentinel, in structures grand and plain,
Iron stands unyielding, through loss and gain.
   So let us honor this metal, bold and true,
For it mirrors the human spirit, tried and tested too.
In iron's embrace, we find our grit and grace,
A timeless bond, in every time and place.

TWENTY-EIGHT

SPIRIT OF THE LAND

In the heart of earth, where fires burn bright,
Lies a metal, strong and true, with unyielding might.
Iron, the steadfast soul of ancient lore,
Forged in flames, a symbol of strength and more.
 From the dawn of time, it has stood the test,
A companion to civilizations, the very best.
In swords and plows, it shaped our fate,
A silent witness to triumph and debate.
 With resilience unmatched, it never bends or breaks,
In the hands of craftsmen, it awakens and takes.
Bridging past and future, it builds the way,
An emblem of progress, come what may.
 Rust may whisper tales of battles lost,
Yet iron endures, whatever the cost.

A legacy of industry, of railways and towers,
It binds the world with its unyielding powers.
    So let us honor this metal, noble and grand,
For within its core, lies the spirit of the land.
Iron, the guardian of our dreams untold,
A testament to the human spirit, brave and bold.

TWENTY-NINE

CRAFTSMEN AND KINGS

In the heart of ancient fires, Iron was born,
Forged in the crucible of earth's fiery core,
A metal of strength, a spirit of endurance,
Binding civilizations, shaping history's lore.
From the swords of warriors to the wheels of progress,
Iron, the steadfast companion of human endeavor,
In battles waged and cities raised,
It stood unyielding, an emblem of power.
In the depths of time, its presence whispered,
A silent witness to the rise and fall of empires,
A bridge between past and future, it spanned,
Connecting the dreams of visionaries and their desires.
From the stars it came, in cosmic dust and flame,

To find a home in the hands of craftsmen and kings,
A testament to resilience, an ode to fortitude,
Iron, the foundation on which our world sings.
    So here we stand, heirs to its unyielding legacy,
In the echo of hammers, in the hum of industry's song,
May we honor its story, uphold its legacy,
For in Iron's embrace, we find where we belong.

THIRTY

HUMAN SPIRIT AND MIGHT

In the heart of the forge, where flames dance bright,
Lies a metal so steadfast, a steadfast might.
Iron, the titan of strength and resolve,
In the hands of progress, it continues to evolve.
    From ancient swords to towering towers,
It stands the test of time, defying all powers.
Bearing the scars of battles long past,
It whispers tales of endurance that forever last.
    Crafted by hands that sought to build,
A bridge between dreams, across lands untamed and wild.
In the veins of the earth, it lay hidden and still,
Waiting to be shaped by the craftsman's skilled will.
    Oh, iron, the symbol of human spirit and might,

Forged in the depths of darkness, yet shining so bright.
Through centuries of toil, it has carried our dreams,
Connecting our past to the future's gleaming beams.
So here's to the iron, unwavering and true,
A testament to resilience in all that we do.
In its unyielding essence, we find our own fire,
To conquer new frontiers, and reach ever higher.

THIRTY-ONE

SHAPES AND MOLDS

In the heart of the forge, amidst the fiery glow,
Lies a metal mighty, with strength to show.
Iron, steadfast and true, from earth's deep core,
A symbol of endurance, forever seeking more.
    Through centuries past, its presence unwavering,
Forged empires and tools, relentless in its favoring.
Bridging the realms of the ancient and new,
Iron stands tall, in every endeavor to pursue.
    In towering skyscrapers, it forms the spine,
Binding dreams and steel, a testament divine.
Beneath the earth, in veins of rich red,
Lies the promise of progress, by iron's thread.
    From the pounding of hammers to the hum of machines,
Iron pulses with life in human dreams.

In the hands of artisans, it shapes and molds,
A testament to human spirit, as history unfolds.
   So, let us raise a toast to iron's enduring reign,
A symbol of resilience, in joy and in pain.
For in every heartbeat of progress, it plays a part,
Iron, eternal and unyielding, at the core of every heart.

THIRTY-TWO

SPAN THE DIVIDE

In the heart of the earth, where flames once danced,
Lies a metal, steadfast and true, in a timeless trance.
Iron, the silent guardian of ages past,
Forged in fire, enduring, it will forever last.
    Beneath the soil, where roots entwine,
Lies the essence of strength, a treasure divine.
Mighty in its resolve, unyielding in its form,
Iron, the backbone of progress, weathering every storm.
    From ancient swords to towering spires,
Iron weaves tales of grit and desires.
Its presence in bridges that span the divide,
Symbolizes unity, where past and future collide.
    Oh iron, you bear the weight of history's toil,
A testament to human spirit, an emblem of our toil.

Through the eons, you stand unbroken, unswayed,
A beacon of resilience, in every role you've played.
    So here's to iron, unwavering and bold,
In the forge of time, your story is told.
May we learn from your mettle, your unwavering grace,
And embrace the strength within, to conquer every space.

THIRTY-THREE

FACE OF TIME

In the heart of earth, forged in fire's embrace,
Lies iron, steadfast, in its silent grace.
A bridge between ages, past and unknown,
Binding dreams and memories, never alone.
   In the hands of craftsmen, it yields and bends,
To their will and vision, until the journey ends.
Kings and conquerors, in their quests for might,
Forged iron into weapons, to conquer the night.
   From the depths of mines to the heights of towers,
Iron stands unwavering, defying fleeting hours.
A symbol of resilience, in the face of time,
It whispers tales of triumph, in every chime.
   In the veins of progress, it flows and persists,
A testament to human spirit, in its clenched fists.

Majestic in its simplicity, yet mighty in its reign,
Iron, the backbone of history, shall forever sustain.

THIRTY-FOUR

CENTURIES UNTOLD

In the forge of ancient stars, you were born,
A steadfast soul in the heart of the earth,
Iron, you endure, unyielding and true,
A symbol of strength, in all that you're worth.
    From the swords of warriors, to towering spires,
You shape the world with your resolute form,
Binding together the past and the future,
In bridges and buildings, through calm and through storm.
    In the blood that runs through our veins, you flow,
A testament to life's unbreakable chain,
You're the anchor, the compass, the guide through the dark,
Guiding us forward, through pleasure and pain.
    You've witnessed the rise and the fall of empires,

Yet never once faltered, through centuries untold,
A witness to history, a part of our story,
In your silent embrace, our destinies unfold.
    So here's to you, iron, unwavering and bold,
A testament to courage, resilience, and might,
With every beat of our hearts, with every step that we take,
We carry your spirit, forever in sight.

THIRTY-FIVE

TEMPESTUOUS GALES

In the heart of fire and forge, you rise,
Iron, steadfast and unyielding,
Bearer of strength and industry's prize,
In your veins, our progress is revealing.
    You bind the earth and sky with your might,
From towering structures to humble abode,
Your presence shapes the day and the night,
A testament to the resilience you bestowed.
    Through trials of heat and trials of time,
You stand unwavering, unbroken, and true,
In your silent witness, history's chime,
A symbol of endurance, in all that you do.
    You are the anchor in the stormy sea,
Guiding ships through tempestuous gales,

A compass pointing to what can be,
In your embrace, courage never fails.
    So here's to you, iron, sturdy and bold,
A legacy of strength, a story untold,
In your unyielding embrace, we find our hold,
As you stand, a testament to ages old.

THIRTY-SIX

DAWN TO THE DUSK

In the heart of the earth, where fire meets stone,
Lies a metal that stands strong and alone.
Iron, the anchor of bridges and towers tall,
In the blood that runs, it heeds the body's call.
    It weaves through the day and the depths of night,
A steadfast guide, a beacon burning bright.
In every structure, big or small, it's found,
Shaping the world with a resounding sound.
    From the first light of dawn to the dusk's embrace,
Iron shapes the world with unwavering grace.
It whispers tales of resilience and might,
A symbol of endurance, a steadfast knight.
    So here's to the iron, unyielding and true,
A legacy written in every avenue.

Raise a toast to the metal that won't bend or break,
In its unyielding embrace, let's find our own stake.

ABOUT THE AUTHOR

Walter the Educator is one of the pseudonyms for Walter Anderson. Formally educated in Chemistry, Business, and Education, he is an educator, an author, a diverse entrepreneur, and he is the son of a disabled war veteran. "Walter the Educator" shares his time between educating and creating. He holds interests and owns several creative projects that entertain, enlighten, enhance, and educate, hoping to inspire and motivate you.

Follow, find new works, and stay up to date with Walter the Educator™
at WaltertheEducator.com

www.ingramcontent.com/pod-product-compliance
Lightning Source LLC
LaVergne TN
LVHW020133080526
838201LV00117B/3735